T0062701

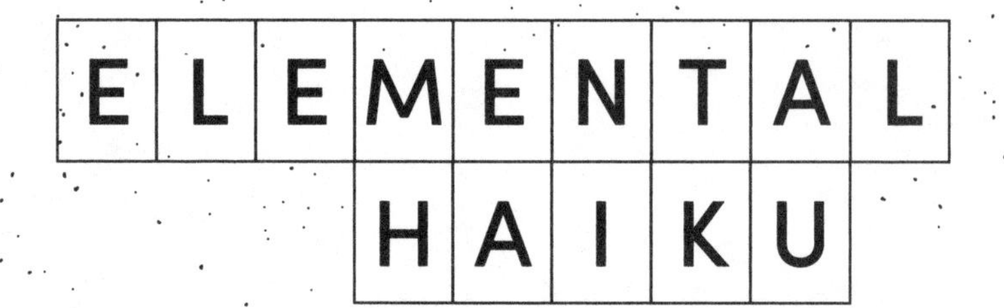

Poems to honor
the periodic table
three lines at a time

MARY SOON LEE

ILLUSTRATIONS BY IRIS GOTTLIEB

TEN SPEED PRESS
California | New York

To my own teacher,
Jane Angliss, from whose lessons
these haiku have grown.

And to all teachers
whose lessons waken a love
of the sciences.

CONTENTS

ACKNOWLEDGMENTS

This book found a home
because of Lisa Rodgers,
my agent and friend.

Lisa Westmoreland's
editorial insight
helped make it better.

As my family—
Andrew, William, and Lucy—
make all things better.

Before this version,
the *Elemental Haiku*
appeared in *Science*.

The illustrations
lending beauty to the words
are Iris Gottlieb's.

And behind the scenes:
Lisa Bieser, Kristi Hein,
David Westmoreland.

INTRODUCTION

I can't remember what age I was when I first learned about the periodic table, but ever since, it has been both familiar and fascinating. In the periodic table, the elements—building blocks of the universe—are marshaled in order of increasing atomic number (the number of protons in their atoms). Then the elements are laid out in rows (known as periods) and columns (known as groups) that reflect underlying atomic structure.

Dmitri Mendeleev, father of the periodic table, devised his first version in 1869. He arranged the elements so that those with similar behavior would line up, leaving gaps in the spots where no known element seemed to belong. A few years later, the first of these missing elements was discovered by Paul-Émile Lecoq de Boisbaudran, who named it gallium. Now, one and a half centuries after Mendeleev's first version, the periodic table contains some 118 known elements, several of them discovered only in the last couple of decades.

One day in December 2016, with no grand plan in mind, I sat down and wrote a haiku for the first of these elements, hydrogen. Hydrogen has an atomic number of one, meaning its atoms contain just one proton, yet hydrogen is part of water and is essential to life:

Hydrogen

Your single proton

fundamental, essential.

Water. Life. Star fuel.

Having written a haiku for hydrogen, I wrote haiku for helium and lithium, the second and third elements in the periodic table. The next day, I came up with haiku for the next several elements, and the project acquired purpose. Progressing in order through the periodic table, I wrote a haiku for each element in turn. Months later, I reached the end of the periodic table, and added one final haiku for element 119, as yet undiscovered as I write this.

This book contains those haiku, together with brief notes on each one. I call them haiku, because I followed a 5-7-5 syllable pattern, and because I tried to capture something of the succinctness, the juxtaposition, the surprise of traditional Japanese haiku. But these are not the tiny, beautifully wrought haiku of Basho or Buson or Issa. There is no cherry blossom to be found, no seasonal references at all. Instead, there are the elements: their chemistry, physics, history.

These *Elemental Haiku* first appeared, without the accompanying notes, in the August 4, 2017, issue of *Science*. They were not the first poetic take on the periodic table, and I hope they won't be the last.

To close this introduction, here is a poem that I wrote as I neared the end of the periodic table. Unlike the haiku themselves, it's a longer, less structured poem, reflecting on my experience writing the elemental haiku, my journey through the periodic table.

The Periodic Table of Elements

I set out in December,
alone, ill-equipped,
but the way clear.

Hydrogen, helium,
the divided couple
of the first row.

Then lithium to neon, the second row,
eight names I'd memorized in high school,
their number, nature, neighbors
well known to me.

Past sodium, magnesium—
that old familiar leap across to aluminum—
from there a quick run through to argon.
The third row done, so soon.
I thirsted for more, for truth,
for primal facts. Firm. Fixed.

Potassium, calcium to the left
of the landmark columns
of transition metals.
The new year turned. I came to zinc,
gallium, germanium, slid on to krypton.
Stood at the end of the fourth row,
caught my breath.

Strangers ahead.

Walked into the eternal expanse
of the fifth row, its rules ordained
by the powers that underpin the universe.
Physics. Chemistry.
The gods.

Rubidium, strontium,
the leaders of the fifth row.
Yttrium, zirconium, niobium, molybdenum.
The milestone of technetium:
first of the radioactive elements,
their atoms randomly transmuting,
but their own original character
constant as the prime numbers,
certain as arithmetic.

January almost complete
before I reached xenon,
the fifth row's right anchor,
noble, steadying, sure.

Row six.
Cesium, barium,
the lanthanides lined up
beneath the main table.
Their electron configurations,
isotopes, melting points, temperaments.
What we learn alters us.
What we choose to learn.

I spent February in the company
of the thirty-two members
of the sixth row.
Rhenium, osmium, iridium.
Platinum's deep treasure.

March now,
the seventh row ahead of me,
the final partial revelations
of humanity's hard-won knowledge.
Elements summoned in laboratories,
nuclear reactors, particle accelerators
by acolytes attempting to answer
that which had never
been answered.

Beneath,
beyond a door we have not opened,
the oracle of the eighth row.

1
H

Hydrogen

Your single proton

fundamental, essential.

Water. Life. Star fuel.

Hydrogen, being the first element in the periodic table, has only a single proton in its nucleus. It is present in water (H_2O) and all known forms of life. Stars such as the Sun fuse hydrogen into helium, releasing energy.

2
He

Helium

Begin universe.

Wait three minutes to enter.

Stay cool. Don't react.

Most helium was formed in the initial minutes after the universe began. Helium is the first element of group 18, the last column of the periodic table. The first six such elements, called the noble gases, are chemically very unreactive. It is not yet known whether the seventh member of group 18, oganesson, is likewise unreactive.

Lithium

Lighter than water,

empower my phone, my car.

Banish depression.

Under standard conditions, lithium is the lightest metal, light enough to float on water. It is widely used in lithium-ion batteries to power everything from watches to electric cars. Lithium carbonate (Li_2CO_3) is one of the medications for bipolar disorder.

Beryllium

Heart of emerald,

your sweetness a toxic trap;

X-rays see through it.

Emeralds are a type of beryl, a mineral composed of beryllium aluminum cyclosilicate ($Be_3Al_2[SiO_3]_6$). Transparent to X-rays, beryllium is used in X-ray tubes. Beryllium compounds taste sweet but are toxic.

Boron

Just doing your job,

holding plant cells together.

No fireworks, no fuss.

Boron is an essential micronutrient for plants, being needed for their cell walls. One of its industrial applications is in flame-retarding compounds added to plastics.

6
C

Carbon

Show-stealing diva,

throw yourself at anyone,

decked out in diamonds.

Carbon will bond with an abundance of other elements, and, like hydrogen, is necessary to all known forms of life. Among the millions of carbon compounds are sugars, proteins, and DNA. Diamond is a crystalline allotrope of carbon (allotropes of an element being the different physical forms it can take).

Nitrogen

Forever cycling

from air, to soil, roots, crops, us.

Exercise addict.

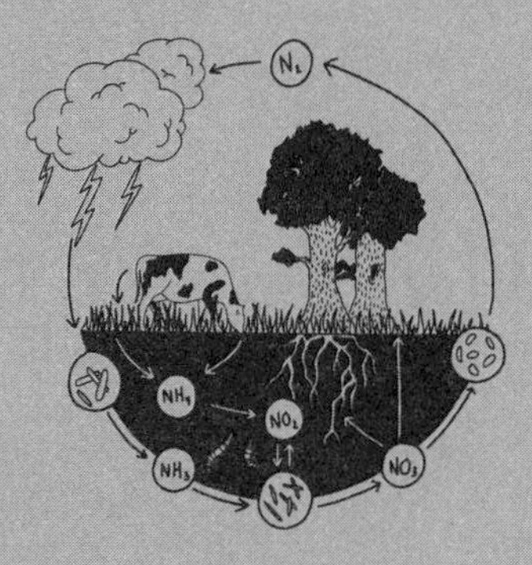

The nitrogen cycle is the complex set of processes by which nitrogen moves between the air, the soil, and living things. Bacteria in the roots of plants such as beans and peas absorb nitrogen from the air and combine it into compounds used both by the plants and by the animals that eat them. Other bacteria convert nitrogen compounds from dead organisms back into nitrogen gas, returning it to the atmosphere. (This brief description of the nitrogen cycle is simplified; for example, nitrogen fixation may be achieved by lightning as well as by microorganisms.)

Oxygen

Most of me is you.

I strive for independence,

fail with every breath.

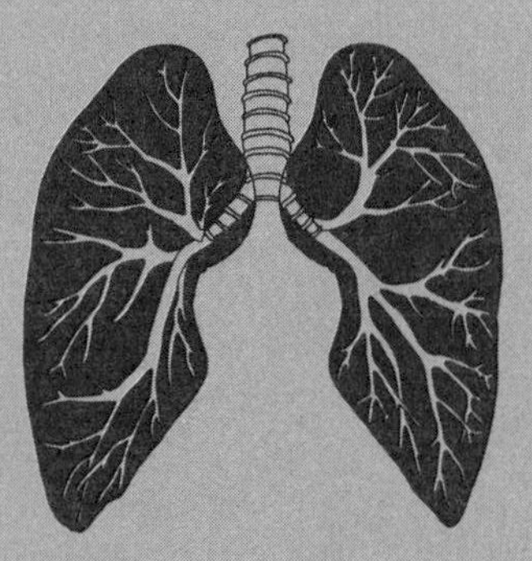

By mass, over half of the human body is oxygen, much of it in the form of water (H_2O). When we breathe, oxygen in the air is absorbed through our lungs into our blood.

9
F

Fluorine

Tantrums? Explosions?

First step: admit the problem.

Electron envy.

Fluorine is the first of the halogens, the elements in group 17 of the periodic table. It is highly reactive; for instance, it reacts explosively with hydrogen. This is because it is the most electronegative element, meaning that its atoms attract electrons most strongly.

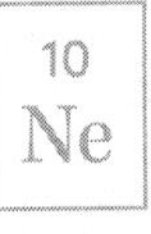

Neon

There's no shame in it.

Advertising pays the bills.

Stop looking so red.

Neon is the second of the noble gases. It is used in neon signs, which ionize neon and produce a bright orange-red glow.

Sodium

Racing to trigger

every kiss, every kind act;

behind every thought.

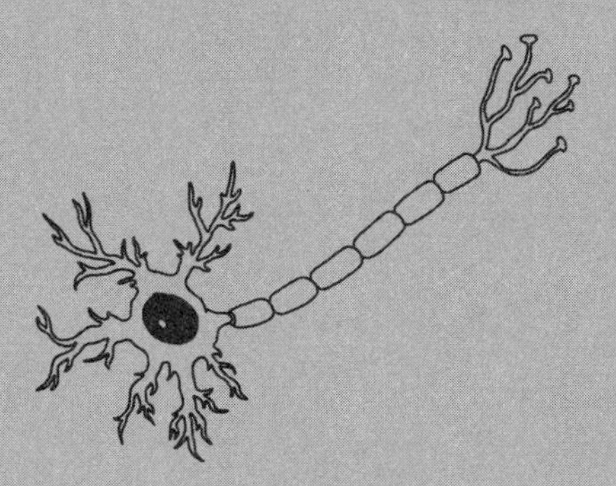

Sodium ions travel across nerve cell membranes, triggering the transmission of nerve impulses and thereby enabling our thoughts and actions.

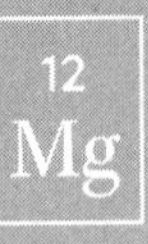

Magnesium

Child of aging stars,

however brightly you burn

they will not return.

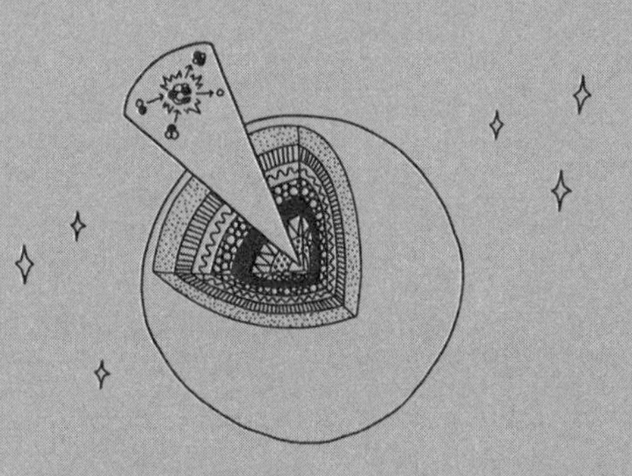

Magnesium burns with a brilliant white flame. It is created in the cores of aging stars by nuclear fusion. More generally, stellar nucleosynthesis is the process by which heavier atomic nuclei are produced in stars from lighter nuclei, by means such as nuclear fusion and neutron capture.

13
Al

Aluminum

This common metal

uncommonly convenient.

Foil. Cans. *Apollo.*

NOTE: I grew up with the British spelling for "aluminum"—"aluminium" (with an extra *i*)—and my first frivolous version of this haiku reflected that:

Aluminium/Aluminum, Al

Spent kindergarten

endlessly writing your name.

One *i* or two *i*'s?

Aluminum is the most common metal and the third most common element in the Earth's crust. Being light, corrosion-resistant, and a good electrical conductor, it is widely used in items from ladders to airplanes. The *Apollo* command and service module had an aluminum honeycomb between sheets of aluminum alloy.

14
Si

Silicon

Locked in rock and sand,

age upon age awaiting

the digital dawn.

Found in rocks and sand, silicon is the second most common element in the Earth's crust. Today we depend on computers and their integrated circuits, which are built on semiconductor wafers, usually made of monocrystalline silicon—hence the term "silicon chip" (and Silicon Valley!).

15
P

Phosphorus

Report, Willie Pete.

Don't hide behind a smoke screen.

How many killed? Maimed?

Willie Pete is a nickname for white phosphorus, an allotrope of phosphorus that burns fiercely. Sometimes employed in smoke grenades, it was used in incendiary weapons in both World Wars, the Vietnam War, and more recent conflicts.

Sulfur

Gunpowder's slaughter.

Fire and brimstone. Damnation.

Penicillin's grace.

As with aluminum, my first version of a haiku for sulfur played on its spelling variants in British and American English:

Sulphur/Sulfur, S

Disrupted first grade,

popping stink bombs, starting fires.

Still can't spell your name.

Although sulfur-free varieties of gunpowder now exist, for centuries sulfur was a critical component. Brimstone is an old name for sulfur. The phrase *fire and brimstone*, found in the King James Version of the Bible, conjured up the torments of hell. On a more positive note, sulfur is one of the chemicals in penicillin.

Chlorine

Low road or high road?

World War I. Gas in trenches.

Or salt shared, tears shed.

Chlorine gas was deployed to terrible effect in the trench warfare of World War I, as was mustard gas, a chlorine compound. Salt, sodium chloride (NaCl), was a valuable trade commodity in the past, and the sharing of bread and salt has symbolic importance in several cultures. Tears contain a small percentage of salt.

18
Ar

Argon

They named you lazy,

but it takes strength to resist.

To stand by. To shield.

Argon, the third of the noble gases, is characteristically unreactive. As a result, it was named after *argos*, the Greek word for *idle* or *inactive*. Due to its inertness, argon serves as a shielding gas in industry and laboratories.

Potassium

Leftmost seat, fourth row,

yearning for the halogens

on the other side.

Potassium is one of the alkali metals, elements in the first (leftmost) column of the periodic table. As such, it has a single valence electron in its outer shell and is highly reactive with the halogens—the elements near the right-hand side of the table in group 17, which are very electronegative (that is, they strongly attract electrons).

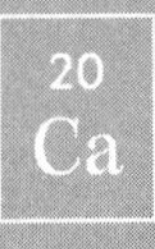

Calcium

The horse's gallop,

the eagle's swiftness both framed

by your quiet strength.

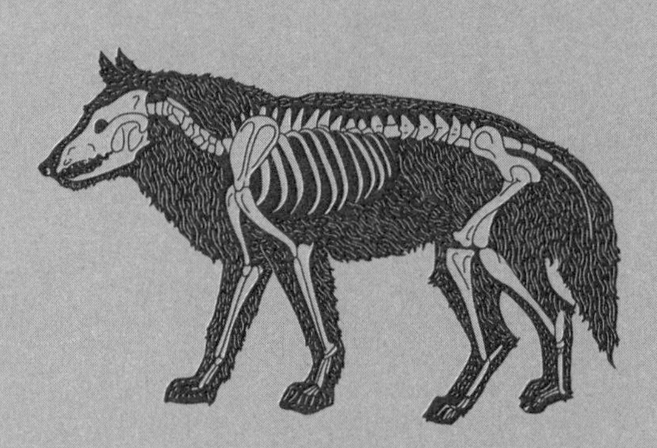

Calcium is a key component in the bones of vertebrates, from hamsters to humans. Hydroxyapatite ($Ca_5[OH][PO_4]_3$), a calcium compound, gives bones compressive strength.

Scandium

How to define you?

Transition metal? Yes? No?

Do labels matter?

Scandium is the first element in the d-block of the periodic table, the middle section of the table containing groups 3 through 12. There are different definitions of *transition metal*, a term that is often applied to all d-block elements, including scandium. However, under other, narrower definitions, scandium is sometimes excluded: if transition metals are defined as elements that can produce a stable ion with an incomplete d subshell (the *d* refers to the particular angular momentum quantum number of that subshell), then scandium doesn't qualify, because the stable Sc^{3+} ion has shed the sole electron in its d subshell.

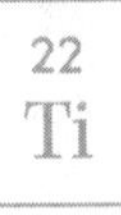

Titanium

Aerospace stalwart,

is the stratosphere enough?

Or only the stars?

Titanium, being both light and strong—not to mention corrosion-resistant—is highly valued in the aerospace sector. From the Boeing 777 and Boeing 787 airplanes to the *Juno* space probe, many aircraft and spacecraft rely on titanium.

Vanadium

Blush with excitement:

lilac, green, blue, and yellow,

shedding electrons.

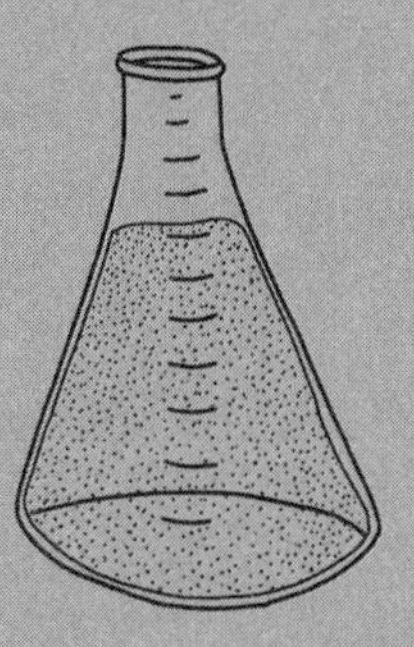

Vanadium can shed varying numbers of electrons to produce different colors in aqueous solution: lilac for the +2 oxidation state, green for +3, blue for +4, and yellow for +5. These colors can be seen—in reverse sequence—by adding zinc to a solution of ammonium metavanadate (NH_4VO_3).

Chromium

Fighting corrosion,

gallant protector of steel,

ready for dragons.

Stainless steel is an alloy of steel and chromium that is more corrosion-resistant than steel. The more chromium is present in stainless steel, the higher its corrosion resistance.

Manganese

In ancient pigments,

seventeen thousand years old,

bulls call from cave walls.

My first version of this haiku relied on the similarity of the words *manganese* and *magnesium*:

Avoid confusion.

In place of your own name, write

"NOT magnesium."

Manganese dioxide (MnO_2) was used as a black pigment in Paleolithic cave art, including the paintings in the Lascaux cave in France, which are estimated to be seventeen thousand years old and which include the astonishing Hall of the Bulls.

26
Fe

Iron

Anvil, axe, nail, plow,

engine, railway, factory.

Servant, friend, partner.

From the Iron Age to the Industrial Revolution, iron has played a huge part in human history.

Cobalt

Traded from Persia,

a blue beautiful as gold,

for a Chinese vase.

Much of the blue underglaze in Ming Dynasty porcelain came from Persian cobalt pigment.

Nickel

Forged in fusion's fire,

flung out from supernovae.

Demoted to coins.

Nickel is created in stars by fusion. Indeed, with the exception of the small amount we've synthesized ourselves, the source of all but the lightest atoms around us is stellar nucleosynthesis. Supernovae explosions scatter such heavier atoms widely. The U.S. Mint makes nickel coins from a cupronickel blend that is one-quarter nickel and three-quarters copper. Dimes, quarters, and dollar coins contain smaller percentages of nickel. Coins from many other countries also contain nickel.

29
Cu

Copper

Before the Bronze Age,

before history began,

bent to the smith's need.

Over ten thousand years ago, people learned to work copper. This predates both the Bronze Age, when people alloyed tin with copper to make bronze, and the invention of writing approximately five thousand years ago.

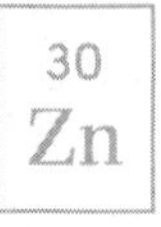

Zinc

Clasp your neighbor tight.

Sound the music while you dance,

trumpet, bugle, horn.

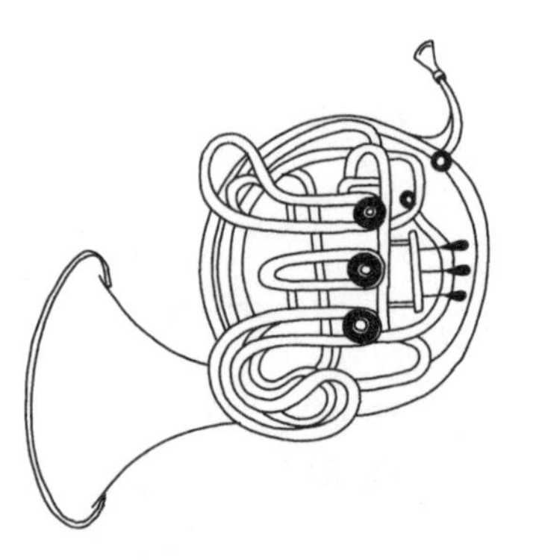

Brass is an alloy of zinc and copper, and copper is zinc's left-hand-side neighbor in the periodic table. Brass instruments include trumpets, bugles, and horns.

Gallium

Melting in my hand,

agent in nuclear bombs,

feigning innocence.

Gallium's melting point of 30°C is low enough that it will melt if held in the hand. In plutonium bombs, plutonium is alloyed with a small percentage of gallium to stabilize it.

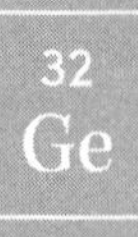

32
Ge

Germanium

Do you miss it still,

the semiconductor crown

that silicon stole?

Germanium was the material of choice for early semiconductor devices such as the first transistors. Today, most integrated circuits are built from silicon rather than germanium.

33
As

Arsenic

Hauled in for questions,

blacklisted, cannot quite shake

your poisonous past.

Many people have been poisoned by arsenic, either deliberately or accidentally. It was referred to as *inheritance powder* due to its abuse as a poison. Accidental cases of arsenic poisoning include one in England in 1900: over six thousand people were sickened by arsenic-contaminated beer, and dozens died.

34
Se

Selenium

So proud to be part

of each drop of formula,

helping babies thrive.

Selenium is essential to humans, though it is toxic in large quantities. Since June 2016, the FDA has required selenium to be included in infant formula.

Bromine

The shame of your name

now lost except to the Greeks.

No need to still fume.

Bromine's name derives from the Greek *bromos* meaning "stench," due to its unpleasant smell. At room temperature, bromine is a fuming, dark-reddish liquid, evaporating readily.

Krypton

For twenty-three years

nobly accepted the quest

to measure meters.

Krypton is the fourth of the noble gases. From 1960 to 1983, in accordance with the International Conference on Weights and Measures, the length of a meter was defined in terms of the wavelength of the orange-red emission line from krypton-86. Since then, the meter is defined as the distance light travels in a vacuum in $\frac{1}{299,792,458}$ of a second.

Rubidium

Temperamental.

Even water starts the flames.

Bunsen's fiery child.

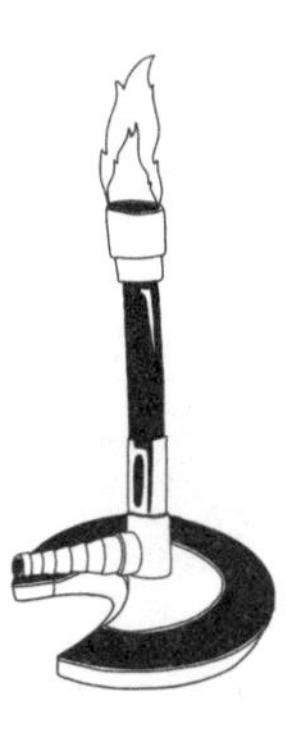

The fourth of the alkali metals, rubidium is highly reactive. For example, its reaction with water is violent, and the resulting hydrogen gas often ignites. Rubidium was jointly discovered in 1861 by Robert Bunsen (inventor of the Bunsen burner, a common piece of laboratory equipment for producing a flame) and Gustav Kirchhoff.

Strontium

Deadly bone seeker

released by Fukushima;

your sweet days long gone.

The radioactive isotope strontium-90 was released after the meltdown of the Fukushima nuclear reactors in 2011 (and also by the 1986 Chernobyl nuclear accident, as well as by earlier nuclear testing). Absorbed by the body like calcium, strontium-90 enters the bones and can lead to cancer. In the nineteenth century, strontium was used to make sugar from sugar beets via the strontian process.

Yttrium

Superconductor
working for minimum wage:
liquid nitrogen.

My first version of this haiku was a commentary on *yttrium*'s unconventional spelling:

That is not a name.
That is a spelling error.
Or a Scrabble bluff.

The first superconductors required such low temperatures that they had to be cooled with liquid helium. In 1987, physicists discovered that an yttrium compound (yttrium barium copper oxide) will exhibit superconductivity at -180°C, warm enough that it can be cooled merely with liquid nitrogen.

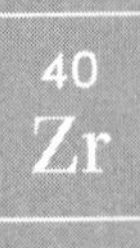

Zirconium

Cosmopolitan,

as at ease in lunar rocks

as in S-type stars.

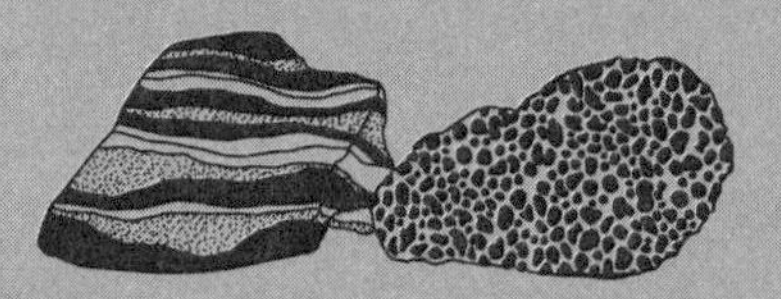

The *Apollo* missions found relatively high concentrations of zirconium in certain lunar rocks. S-type stars are comparatively cool stars characterized by having pronounced zirconium oxide (ZrO) bands in their spectra.

Niobium

Are you still pining

for the name columbium,

its lost poetry?

Columbia was a poetic name for America. In 1801, Charles Hatchett discovered what he believed to be a new element and called it *columbium.* Subsequently people thought Hatchett's columbium might be just a form of the element tantalum. In the 1840s, Heinrich Rose identified a second element in tantalum ores and named it *niobium.* Hatchett's columbium and Rose's niobium are one and the same, and in 1949, niobium became the element's official name.

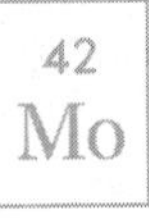

Molybdenum

Mistaken for lead—

such injustice! Not deadly,

but vital for life.

The name molybdenum comes from *molybdos*, the Greek word for "lead," because for a long time molybdenite, one of its ores, was mistaken for lead ore or graphite. While lead poisoning is an ongoing health concern, molybdenum is an essential trace element for humans.

Technetium

Unstable, short-lived,

yet found within red giants:

birthed in their bellies.

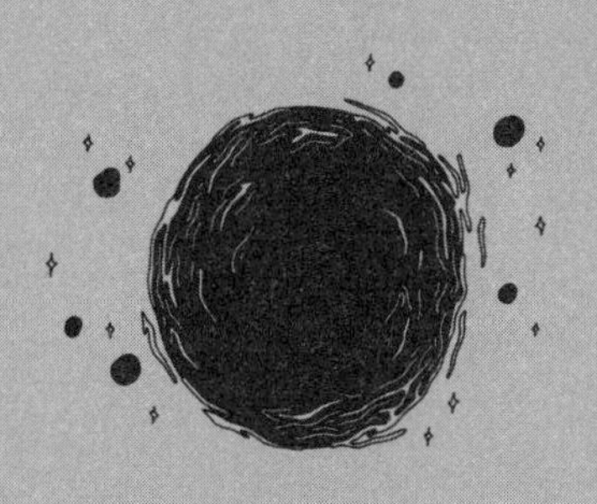

Technetium is radioactive, the half-lives of its isotopes being four million years or less ("short-lived" is a relative term in this haiku!). The half-life of a radioactive isotope is the average time needed for half its atoms to undergo radioactive decay. When technetium was detected in the spectra of red giants by Paul Merrill in 1952, its comparatively short life implied that it must have been synthesized inside the stars.

44
Ru

Ruthenium

Just below iron,

letting him hog the spotlight,

oh noble metal.

Noble metals are those that are highly corrosion-resistant, and ruthenium is such a metal. It is positioned directly beneath iron in the periodic table.

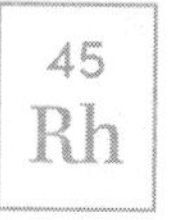

Rhodium

Battling pollution

by transforming exhaust gas,

catalyst for change.

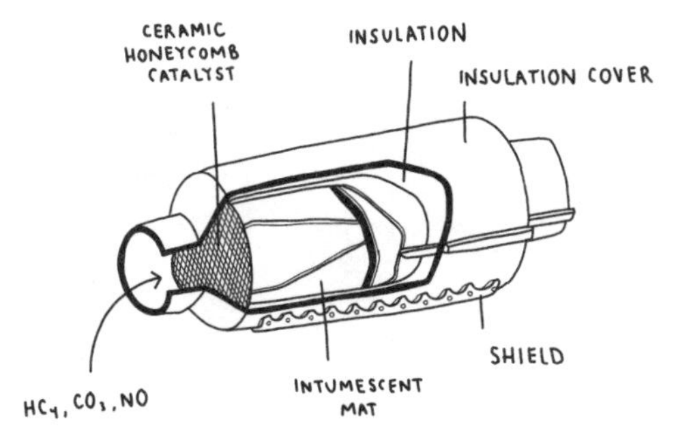

Rhodium is a catalyst in three-way catalytic converters, which make car emissions less harmful.

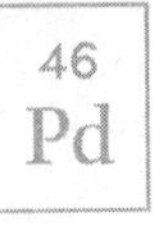

Palladium

How much hydrogen

must you swallow up to please

cold fusion dreamers?

Palladium has a remarkable capacity for absorbing hydrogen gas—it can absorb nine hundred times its own volume of hydrogen. This ability has led people to dream of using palladium to facilitate cold fusion.

Silver

Treacherous treasure,

avarice tarnishing us,

photos claiming souls.

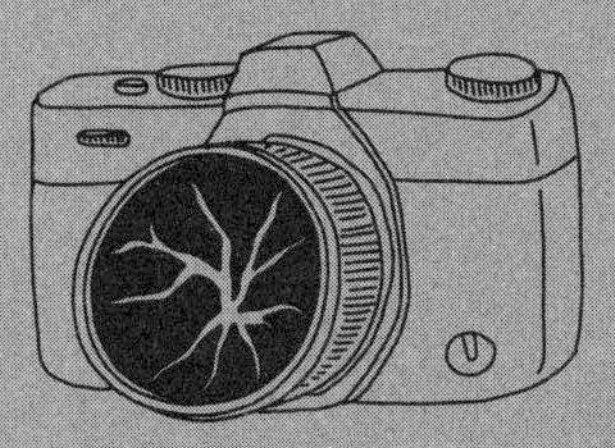

Silver has long been treasured, even though it tarnishes. Silver halides, which are light-sensitive, are used in photographic film. Reportedly, some cultures believed that having their picture taken might harm them or steal their souls.

Cadmium

Defined the angstrom

until krypton ousted you.

Now looking for work.

For decades, the angstrom, a tiny unit of length, was defined in terms of the wavelength of the red emission line from cadmium. In 1960, the angstrom was redefined as one ten-billionth of a meter, with the meter itself being redefined in terms of krypton.

49
In

Indium

Cadmium's daughter

by way of neutron capture;

adrift in stardust.

One example of stellar nucleosynthesis—the process by which heavier atomic nuclei are produced in stars from lighter nuclei—is the creation of indium-115. First, cadmium-115 is formed from cadmium-110 by successive neutron capture; then beta decay (a type of radioactive decay) of cadmium-115 forms indium-115.

Tin

Your magic number

a secret no one could guess

back in the Bronze Age.

For thousands of years—indeed, since the start of the Bronze Age—tin has been alloyed with copper to make bronze. In nuclear physics, the numbers 2, 8, 20, 28, 50, 82, and 126 are called magic numbers. Nuclei that contain a magic number of protons—or indeed of neutrons—are more stable due to having filled nuclear shells. Tin, with fifty protons, is an example.

Antimony

Outline eyes with kohl

in the ancient tradition.

Forego explosions.

The ancient Egyptians made kohl, a type of eyeliner, from stibnite, an antimony ore. One form of antimony is named *explosive antimony* for its propensity to explode.

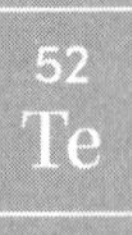

Tellurium

So rare here on Earth,

because you fled into space.

Sighted by Hubble.

Tellurium is among the ten rarest elements found in the Earth's crust, yet it is more abundant in the wider universe. Much of the tellurium that would otherwise have been on Earth is thought to have dispersed into space during the planet's formation or early history. The Hubble Space Telescope detected tellurium in ancient stars by looking at their spectra.

53
I

Iodine

A fortunate fluke:

found fuming from French seaweed,

sublime purple gas.

Iodine, the fourth member of the halogens, sublimes to produce a violet-colored gas. It was discovered by Bernard Courtois in 1811 as he was processing seaweed ash. When Courtois added too much sulfuric acid, he released a cloud of purple gas: iodine.

54
Xe

Xenon

First noble compound?

Hexafluoroplatinate?

Making history.

Xenon is the fifth of the noble gases, the elements in the rightmost column of the periodic table (known as group 18) that are very unreactive. In 1962, Neil Bartlett made chemistry history by announcing he'd succeeded in making the first noble-gas compound by reacting xenon with platinum hexafluoride. He identified the compound as xenon hexafluoroplatinate ($Xe[PtF_6]$).

Cesium

Hot-headed firebrand
whose violent behavior
hides a softer side.

Cesium, the fifth of the alkali metals, is extremely reactive. When dropped into water, it explodes violently, and it is pyrophoric, meaning that it ignites spontaneously when exposed to air. Yet it is also a very soft metal, softer than lead or gold on the Mohs scale.

Barium

Let those enduring

your enemas remember

fireworks' bright splendor.

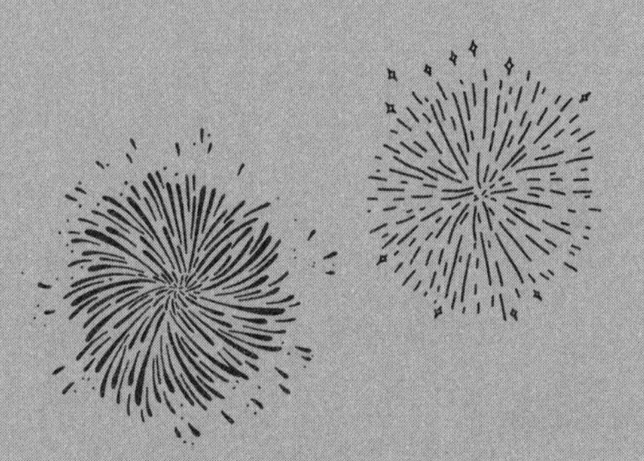

During a barium enema, a medical procedure to inspect the colon, barium sulfate solution is introduced rectally and an X-ray is taken. Barium compounds give fireworks a green color.

57
La

Lanthanum

Down in the cheap seats

underneath the main table,

leading your namesakes.

For convenience, the periodic table is normally drawn with two rows of elements beneath the main part of the table. (If these two rows are inserted into their correct place in the table, it becomes very wide.) Lanthanum is the first element of the first of these rows, a row the elements of which are called the lanthanides.

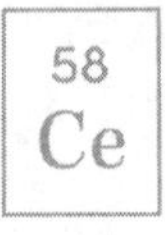

Cerium

More common than lead,

and yet an ardent member

of the rare-earth club.

The International Union of Pure and Applied Chemistry (IUPAC) defines the rare-earth elements as the fifteen lanthanides plus yttrium and scandium. Cerium, being the second of the lanthanides, is thus a rare-earth element. It is the most abundant of the rare-earth elements in the Earth's crust and is indeed more common than lead. (When the first rare-earth elements were discovered, they were thought to be rarer than later proved to be the case; hence their name.)

Praseodymium

Magnetic cooling.

Absolute zero beckons.

Approach the limit.

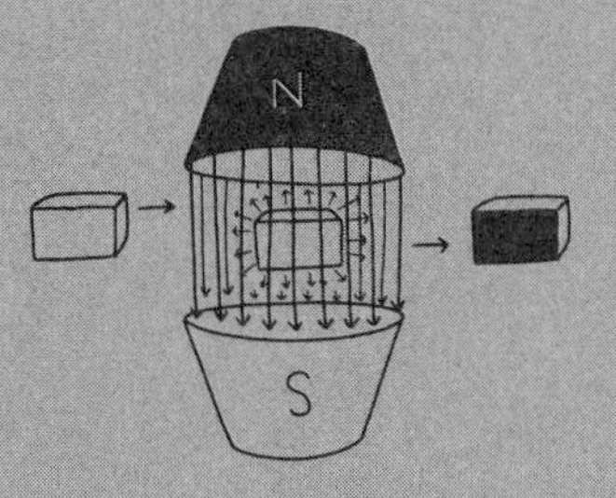

There's a praseodymium-nickel alloy that is highly magnetocaloric, meaning that its temperature changes in response to the changing strength of a magnetic field applied to it. This has enabled scientists to get very close to absolute zero: within one-thousandth of a degree Celsius.

Neodymium

Delved from China's mines,

mixed with iron and boron.

Strong magnets, high costs.

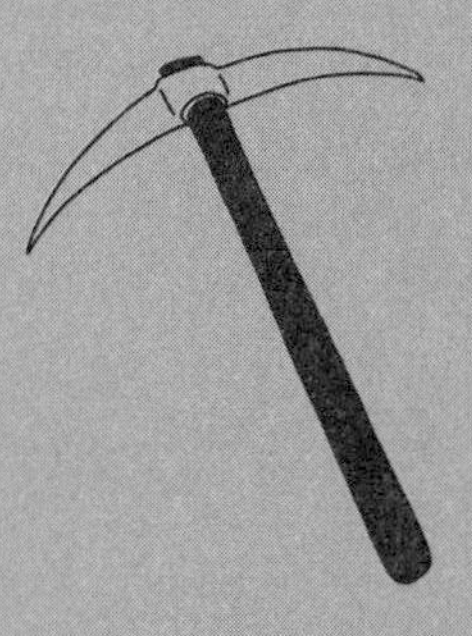

Extremely powerful magnets can be made from an alloy of neodymium, iron, and boron. Much of the supply of neodymium comes from China, and unfortunately, mining has caused serious environmental consequences, including a severely polluted lake.

Promethium

Instability,

radioactivity;

lanthanide bad boy.

Promethium is radioactive. It is the fifth of the lanthanide elements, and the only one with no long-lived isotope, its isotopes all having half-lives of under twenty years.

Samarium

Bone deep, such kindness;

curtailing cancerous cells,

easing others' pain.

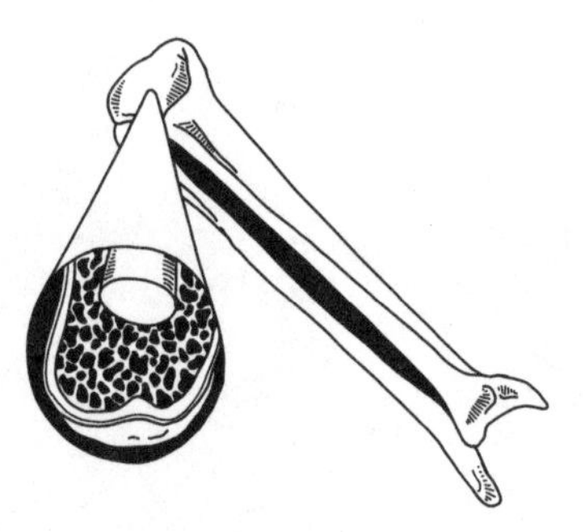

Samarium (^{153}Sm) lexidronam ($C_6H_{17}N_2O_{12}P_4Sm$), containing the radioactive isotope samarium-153, is given to patients to reduce the pain of bone cancer.

Europium

Defending euros,

contained in every banknote,

glowing in UV.

Fittingly, the element named for Europe is included in euro banknotes to help prevent counterfeiting. Euros incorporate europium oxide (Eu_2O_3), which fluoresces red under ultraviolet light.

Gadolinium

Neutron devourer,

let us harness your hunger

within reactors.

One of gadolinium's isotopes, gadolinium-157, is the best neutron absorber among the stable nuclides. As such, gadolinium is employed in nuclear reactors. Xenon-135 is an even better neutron absorber, but it is radioactive.

Terbium

Join with your neighbor

to pulse to the magnet's beat.

Expand. Contract. Dance.

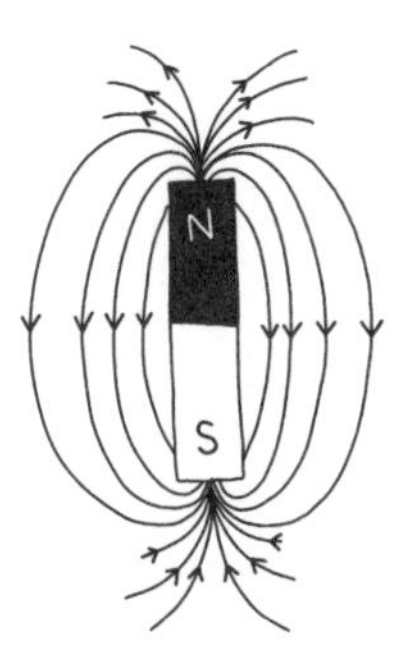

Terfenol-D, an alloy of terbium with its neighboring element dysprosium and with iron, is magnetostrictive, meaning that it expands or contracts in response to a changing magnetic field.

Dysprosium

Your discovery

such a painstaking process,

playing hard to get.

Dysprosium was named after the Greek word *dysprositos*, meaning "hard to get at." Its discoverer, Paul-Émile Lecoq de Boisbaudran, chose the name because he had to go through a very laborious process to separate its oxide.

Holmium

The root of the name

elementary, my dear.

Stockholm, not Sherlock.

The discovery of both holmium and thulium (another of the lanthanides) is credited to Per Teodor Cleve, a Swedish chemistry professor. He named holmium after Holmia, the Latin word for Stockholm, capital of Sweden.

Erbium

Internet helpmate,

improving fiber optics.

Rose-tinted data?

Fiber optics is one of the primary means for transmitting data, including data associated with the internet. Erbium-doped fiber amplifiers (EDFAs) boost fiber-optic signals. In the form of erbium oxide (Er_2O_3), erbium tints glass pink.

Thulium

Fifteen thousand steps

to gain the first pure sample.

Extreme chemistry.

As mentioned in the note on holmium, Per Teodor Cleve discovered thulium as well as holmium. However, it took about thirty more years before Charles James was able to separate a pure sample of thulium (as thulium bromate, $Tm[BrO_3]_3$), an effort which required some fifteen thousand recrystallizations!

Ytterbium

Stable to within

two parts in a quintillion:

atomic clock ticks.

In 2016, the National Institute of Standards and Technology reported that, by combining two ytterbium lattice clocks, they had built an atomic clock that was stable to a remarkable 1.5 parts in a quintillion.

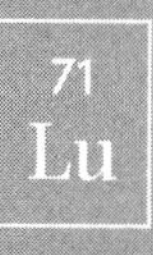

Lutetium

Helicopter dad,

keeping your electrons close

to your nucleus.

Lutetium is the last of the lanthanides. Due to a phenomenon called the lanthanide contraction, its atoms and ions are smaller than would otherwise be the case.

Hafnium

Yield the holy grail:

induced gamma emission.

Let the price be low.

One of the nuclear isomers of hafnium controversially attracted attention as the possible basis for a weapon, whereby energy stored in the isomer would be released in a burst of gamma rays (induced gamma emission). In 2008, scientists at Lawrence Livermore National Laboratory concluded that such a scheme was impractical, and that the cost of producing the isomer would probably be prohibitively high.

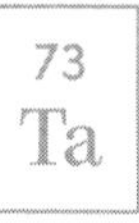

Tantalum

Smaller and lighter,

your tiny capacitors,

digital darlings.

Tantalum-based capacitors are small and light, and hence widely used in digital electronics.

Tungsten

Illuminated

the twentieth century,

glowing with such pride.

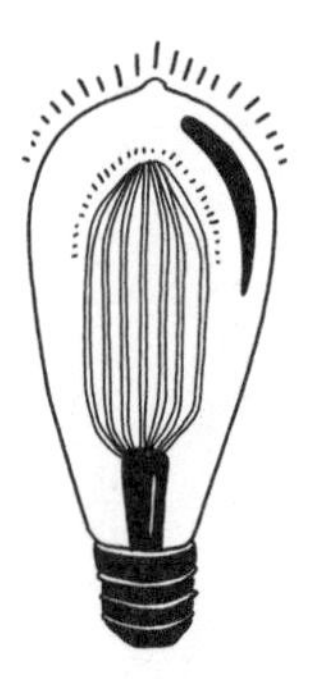

Tungsten-filament light bulbs were long the default choice, although now they have largely yielded to more energy-efficient alternatives.

Rhenium

1925.

Found: new element. Stable.

The last to be caught.

This haiku takes 1925 as the discovery date for rhenium; in that year a German team successfully isolated and identified it, making it the last stable element to be discovered. I note that Masataka Ogawa probably separated rhenium earlier, in 1908, but he thought he had found the forty-third element in the periodic table, rather than the seventy-fifth. (Rhenium has atomic number 75.)

Osmium

Humiliated

underneath the dunce's cap,

densest in the class.

Osmium is the densest of the elements, or at least the densest of those for which a density is known. The density of some of the synthetic elements is not yet known.

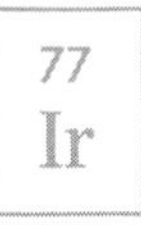

Iridium

The dinosaurs gone:

your fingerprint in the clay.

Incriminating.

The mass extinction of the dinosaurs, about sixty-five million years ago, is thought to be due to a large meteor impact. There's an important piece of evidence for this: the layer of rock dating from that era contains markedly higher levels of iridium than is usual in the Earth's crust. Meteoroids similarly contain relatively high levels of iridium.

78
Pt

Platinum

In a basement vault,

nestled within two bell jars,

the measure of mass.

For over a century, the kilogram was defined as the weight of a particular cylinder of metal—the international prototype kilogram—an alloy of 90 percent platinum and 10 percent iridium, which was carefully stored in a basement vault in Paris. In May 2019, the kilogram was redefined in terms of Planck's constant, by decreeing that constant to be exactly $6.62607015 \times 10^{-34}$ $kg \cdot m^2 \cdot s^{-1}$.

Gold

Deep they delved for thee,

yet deeper still thy dwelling

in the Earth's dark core.

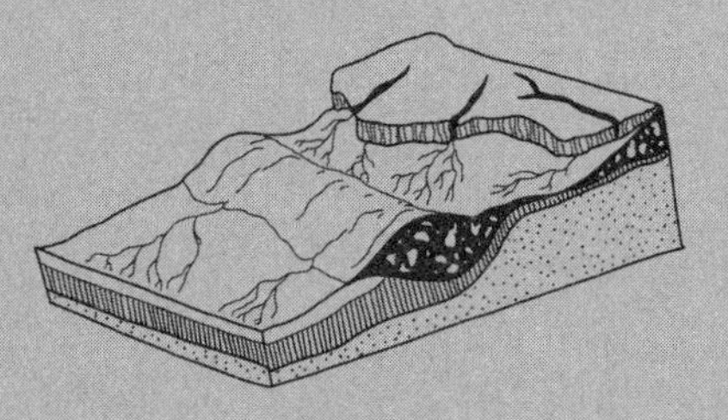

Gold has been sought after by humans for millennia. The deepest gold mines extend over two miles underground. Far beneath the deepest of those mines, it is believed that the Earth's core contains not only iron, but also gold and platinum.

Mercury

Madness the price paid

for your molten alchemy.

Metal. Planet. God.

Mercury is a silvery metal notable for being liquid under standard conditions. It featured in alchemical traditions in different parts of the world. Toxic, mercury damages the central nervous system and can cause madness. Mercury is also the name of both a Roman god and its namesake planet.

81
Tl

Thallium

As a sulfate salt,

handy for killing rodents

or one's relatives.

Thallium sulfate (Tl_2SO_4) was once a readily available rat poison. Alas, it is also lethal to humans. In Australia in the 1950s, a number of murders and attempted murders were perpetrated with thallium sulfate, including the case of a grandmother who poisoned several family members with it.

Lead

Lecherous plumber

with an appetite for both

acids and bases.

The word plumber is derived from *plumbum*, the Latin for lead. Plumbers at least as far back as the Romans favored lead for their pipes. Lead reacts with both acids and bases.

Bismuth

Given a half-life
of ten to the nineteen years,
de facto stable.

In 2003, French researchers found that bismuth-209, once believed to be a stable isotope, has a half-life of 19,000,000,000,000,000,000 years.

Polonium

Hidden in pitchblende,

gleaned by Marie and Pierre,

their radiant child.

Marie and Pierre Curie arduously separated two previously unknown elements from pitchblende, announcing the discovery of polonium in July 1898 and of radium in December 1898. Polonium-210 is intensely radioactive, glowing as it excites the surrounding air.

Astatine

Naturally scarce.

Perhaps an ounce to be had

in the whole Earth's crust.

The longest-lived isotope of astatine has a half-life of under nine hours. Astatine arises naturally in the Earth's crust from the decay of other radioactive elements and then rapidly decays itself, so at any given moment very little is present.

Radon

Homeowners' hazard,

skulking down in the basement,

plotting your decay.

Radon gas is heavier than air and can build up in homes, posing a health hazard due to its radioactivity.

Francium

Last seat, first column,

there maybe twenty minutes

before you vanish.

Francium is the bottom element in the first column of the periodic table. Its longest-lived isotope, francium-223, has a half-life of twenty-two minutes.

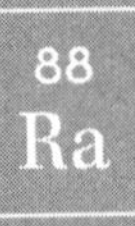

Radium

Licked by the women

painting luminous watches.

How much time stolen?

Paint containing radium was once used to make luminous watches. In the 1920s, a group of five women, known as the Radium Girls, sued their employer for illness caused by licking their brushes as they painted such watches. Tragically, all five women—plus many of their coworkers—subsequently died from radium poisoning.

89
Ac

Actinium

Captain of the team

banished to the bottom bench,

troublemakers all.

The actinides are the elements in the second of the two rows that are normally drawn below the main part of the periodic table. The first of the actinides, and the one for which they are collectively named, is actinium. All the actinides are radioactive.

Thorium

Electrons display

relativistic effects

in the table's depths.

Being an actinide, thorium is in the bottommost section of the periodic table. Heavier elements, such as thorium, experience relativistic effects on electrons; that is, their behavior is noticeably influenced in accordance with Einstein's theory of relativity. In the case of another element, gold, this is quite conspicuous: gold's color is due to relativistic effects.

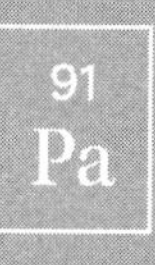

Protactinium

Both of your neighbors
proffer nuclear power;
you clog reactors.

Protactinium's neighbors in the periodic table—thorium and uranium—have both been used to fuel nuclear power reactors. Protactinium-233 is an intermediate product in thorium-based reactors; however, being a strong neutron absorber, it slows down the nuclear reactions.

92
U

Uranium

Manhattan Project.

The elephant in the room,

never forgetting.

On August 6, 1945, Little Boy, a uranium-based nuclear bomb developed by the Manhattan Project, was dropped on the Japanese city of Hiroshima with devastating effect. It was the first nuclear weapon exploded in warfare.

Neptunium

One small step toward

Pioneer and *Voyager*,

space ambassadors.

The *Pioneer* and *Voyager* space probes are powered by plutonium-238 radioisotope thermoelectric generators, and the plutonium-238 itself was synthesized from neptunium. Other notable space probes, including *Galileo* and *Cassini*, likewise obtained power from plutonium-238 manufactured from neptunium.

Plutonium

Manhattan Project.

Climbing the tree of knowledge

for forbidden fruit.

The haiku for uranium and plutonium share the same opening line, both elements branded by their association with the Manhattan Project, the World War II effort that developed nuclear weapons. On August 9, 1945, three days after the detonation of the uranium-based Little Boy, a plutonium-based nuclear bomb, nicknamed Fat Man, was dropped on the Japanese city of Nagasaki. Genesis 2:17 of the Bible begins, "But of the tree of the knowledge of good and evil, thou shalt not eat of it."

Americium

Alpha particles
dispatched in smoke detectors
to protect and serve.

The isotope americium-241 is common in smoke detectors. It emits alpha particles, ionizing the air and increasing its conductivity. If smoke is present, conductivity decreases and the alarm sounds. The motto of the Los Angeles Police Department is "To Protect and to Serve."

Curium

Crew member assigned

to spectrometer duty

on the Mars rovers.

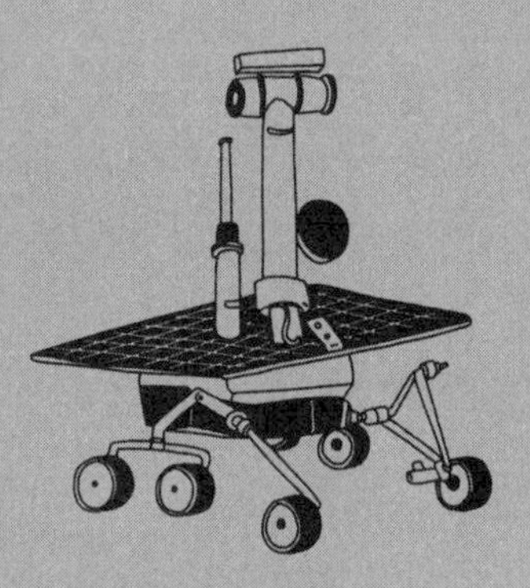

The spectrometers on the Mars *Pathfinder*, *Spirit*, *Opportunity*, and *Curiosity* rovers all relied on curium-244 as a source of alpha particles to bombard the rocks under investigation.

97
Bk

Berkelium

Face the firing squad.
Yield exotic elements
under bombardment.

Most of the elements toward the end of the periodic table were found by deliberately synthesizing them from lighter elements. Berkelium may be synthesized from americium or curium, which appear just before it in the periodic table. In turn, heavier elements may be synthesized from berkelium. Californium, the next element in the periodic table, may be produced by bombarding berkelium with neutrons. Tennessine, which is currently the penultimate element in the periodic table, was first synthesized by bombarding berkelium with calcium ions.

Californium

Freely fissioning,

spitting out enough neutrons

to start reactors.

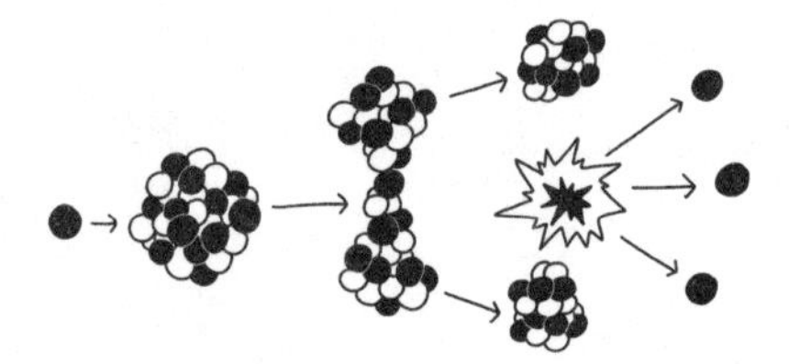

Californium-252 is a powerful neutron emitter, its nuclei spontaneously fissioning into lighter elements with an accompanying release of neutrons. Indeed, it emits sufficient neutrons that it can start up nuclear reactors.

Einsteinium

Forged in the furnace

of $E = mc^2$.

Ivy Mike's baby.

Einsteinium was discovered in the fallout from the first successful test of a hydrogen bomb in 1952. The code name for this project was Ivy Mike. Einsteinium was named after Albert Einstein, famous for his equation $E = mc^2$. This equation describes the relationship between energy and mass that underlies the power of nuclear bombs, including that of Ivy Mike.

Fermium

Hydrogen bomb test

yields collateral science,

fallout bonanza.

Like einsteinium, fermium was first discovered in the fallout from the hydrogen bomb test code-named Ivy Mike. Thus two new elements were discovered in the wake of this bomb test.

Mendelevium

Cyclotron harvest

named for the table's father:

seventeen atoms.

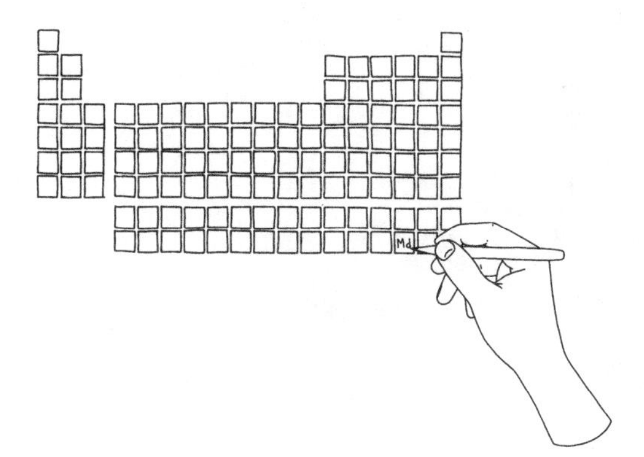

Mendelevium is named after Dmitri Mendeleev, father of the periodic table. The element was first synthesized in 1955, when a grand total of seventeen atoms of mendelevium were made by bombarding einsteinium with helium ions in a cyclotron. Scientists will go to great lengths to synthesize even a few atoms of a new element.

Nobelium

No prize to award

for the decades disputing

your discovery.

Nobelium is named after Alfred Nobel, who bequeathed most of his fortune to create the Nobel Prizes. Teams of scientists in Sweden, the United States, and the Soviet Union all claimed to have discovered this element, the 102nd in the periodic table. It was decades before the dispute was finally settled. In the end, the team from Dubna, Russia, was officially credited with the discovery.

Lawrencium

Silvery, we think,

guessing, not having seen you

with the naked eye.

Lawrencium—along with the other elements near the end of the periodic table—has only been produced in tiny quantities, too small for us to see. It is conjectured that it would look similar to lutetium, the lanthanide directly above it in the periodic table, which is silvery white.

Rutherfordium

No known isotope

condescends to hang around

much more than an hour.

Rutherfordium's longest-lived known isotope, rutherfordium-267, has a half-life of 1.3 hours.

Dubnium

A silent soldier

in the Transfermium Wars.

Whose side were you on?

The transfermium elements are those appearing after fermium in the periodic table. Several of these elements, including dubnium, were the subjects of a bitter years-long naming controversy, dubbed the Transfermium Wars.

Seaborgium

Out, out, brief candle!

Mere minutes upon the stage.

Strut, fret, and exit.

In a famous speech, Shakespeare's Macbeth says, "Out, out, brief candle! Life's but a walking shadow, a poor player that struts and frets his hour upon the stage, and then is heard no more." The most stable isotope of seaborgium yet discovered has a half-life of a few minutes.

Bohrium

High-speed chemistry:

Forming your oxychloride

before you vanish.

Bohrium is another radioactive element that decays very quickly. Despite this, in 2000, a team of scientists succeeded in making its oxychloride (BhO_3Cl) from six atoms of bohrium.

Hassium

Historic hassle,

crafting sodium hassate,

atom by atom.

Hassium's most stable known isotope has a half-life of only ten seconds. Nonetheless, in the early 2000s, scientists in Darmstadt, Germany, formed sodium hassate ($Na_2[HsO_4(OH)_2]$) from a few atoms of hassium. (I note that hassium tetroxide [HsO_4] was formed slightly earlier, but I selected the hassate so that I could pair "hassle" with "hassate.")

Meitnerium

Merely conjectured,

your chemical properties.

No data at all.

At the time I write this, there have been no reported experiments into the chemistry of meitnerium, an element first synthesized in the 1980s.

Darmstadtium

Nickel ions hurled

by the billions of billions

to score the first goal.

The discovery of darmstadtium is credited to scientists in the German city of Darmstadt, after which the element is named. In 1994, they bombarded a lead target with over 1,000,000,000,000,000,000 nickel ions to create the first atom of darmstadtium.

Roentgenium

In the group of coins:

copper, silver, gold, and you.

Untried, worth unknown.

Roentgenium is in group 11 of the periodic table; that is, in the eleventh column of the table. The other elements in the group are copper, silver, and gold, metals that have all been used in coins. As yet, no experiments investigating roentgenium's chemistry have been reported, although it is thought that it will be similar to gold and silver.

Copernicium

Electrons arrayed
in atomic orbitals
as if round a sun.

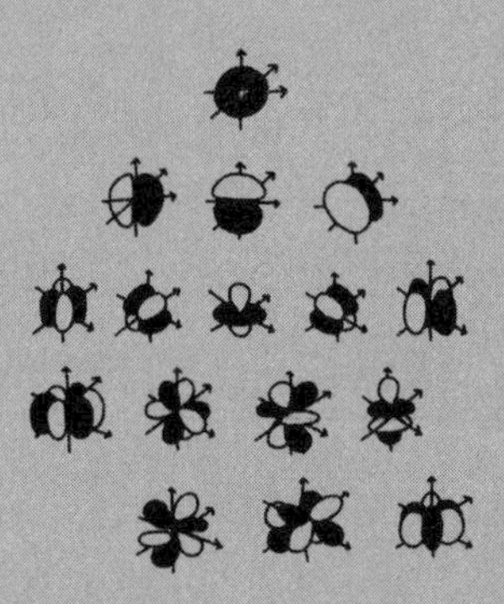

Copernicium is named after Nicolaus Copernicus, who proposed the idea that the Earth and other planets revolve around the Sun. The haiku is an homage to Copernicus's ideas, but atomic orbitals are *not* close analogues to planetary orbits; rather, they describe the region around a nucleus where an electron is likely to be found. Atomic orbitals are defined by wave functions. Note: The notion that the Earth orbits the Sun had been proposed more than a millennium before Copernicus, by Aristarchus of Samos.

Nihonium

A stable island?

Might heavier isotopes

linger for longer?

It is conjectured that there may be a so-called island of stability for isotopes of heavy elements. Isotopes with appropriate numbers of protons and neutrons would be on this island and would have half-lives that are longer (perhaps even far longer) than otherwise typical for elements of such a high atomic number. In the case of nihonium, its lightest known isotopes have half-lives of less than a tenth of a second, whereas the heaviest known isotopes have half-lives over a second. These heavier isotopes may lie closer to the island of stability.

Flerovium

Mirage? Mistake? Truth?

Would further neutrons lend you

a doubled magic?

The notes for nihonium on page 119 discuss the island of stability. It is an open question where the center of the island of stability would be, if the island exists at all. One of the possible contenders for the center has been at 114 protons and 184 neutrons—that is, the isotope flerovium-298, an isotope that has not yet been synthesized. (The isotopes of flerovium that have been successfully synthesized have fewer than 184 neutrons.)

Moscovium

First fashioned by us

in this new millennium,

future uncharted.

Moscovium was first synthesized in 2003 at the Joint Institute of Nuclear Research in Dubna, Russia.

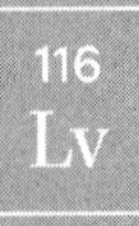

Livermorium

Nearing the end now.

the antepenultimate

seventh-row inmate.

Livermorium is the third-to-last element in the seventh row of the periodic table.

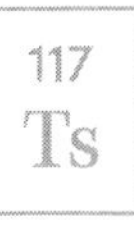

Tennessine

Our most recent find,

evanescent halogen.

New kid on the block.

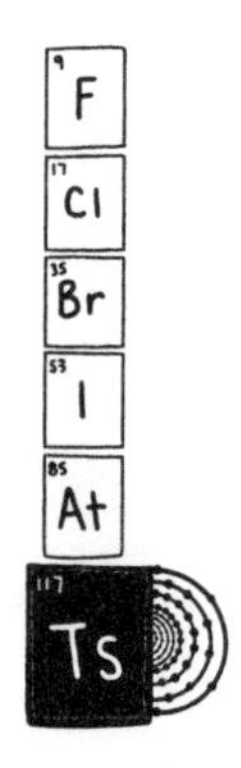

Tennessine is in group 17 of the periodic table, the halogen group. The isotopes of tennessine thus far produced are very unstable, quickly decaying. First synthesized in 2010, tennessine is the most recently discovered element.

Oganesson

The end of the line,

your millisecond half-life

brings down the curtain.

Oganesson is the heaviest element yet synthesized. Oganesson-294, the first of its isotopes to be synthesized, has a half-life of about one millisecond.

Ununennium

Will the curtain rise?

Will you open the eighth act?

Claim the center stage?

At the time of this writing, no one has reported synthesizing element 119, which would be the first member of the eighth row of the periodic table. For now, the element has the temporary name of ununennium, but its discoverers will have the honor of proposing a permanent name.

THE PERIODIC TABLE OF ELEMENTS

1 H								
3 Li	4 Be							
11 Na	12 Mg							
19 K	20 Ca	21 Sc	22 Ti	23 V	24 Cr	25 Mn	26 Fe	27 Co
37 Rb	38 Sr	39 Y	40 Zr	41 Nb	42 Mo	43 Tc	44 Ru	45 Rh
55 Cs	56 Ba		72 Hf	73 Ta	74 W	75 Re	76 Os	77 Ir
87 Fr	88 Ra		104 Rf	105 Db	106 Sg	107 Bh	108 Hs	109 Mt
119 Uue								

57 La	58 Ce	59 Pr	60 Nd	61 Pm	62 Sm
89 Ac	90 Th	91 Pa	92 U	93 Np	94 Pu

									2 He
			5 B	6 C	7 N	8 O	9 F	10 Ne	
			13 Al	14 Si	15 P	16 S	17 Cl	18 Ar	
28 Ni	29 Cu	30 Zn	31 Ga	32 Ge	33 As	34 Se	35 Br	36 Kr	
46 Pd	47 Ag	48 Cd	49 In	50 Sn	51 Sb	52 Te	53 I	54 Xe	
78 Pt	79 Au	80 Hg	81 Tl	82 Pb	83 Bi	84 Po	85 At	86 Rn	
110 Ds	111 Rg	112 Cn	113 Nh	114 Fl	115 Mc	116 Lv	117 Ts	118 Og	

63 Eu	64 Gd	65 Tb	66 Dy	67 Ho	68 Er	69 Tm	70 Yb	71 Lu
95 Am	96 Cm	97 Bk	98 Cf	99 Es	100 Fm	101 Md	102 No	103 Lr

SELECTED BIBLIOGRAPHY

This is a selective, not a comprehensive, bibliography. In addition, I browsed the internet extensively. Wikipedia and the various NASA websites proved especially helpful, but other online sources that I used repeatedly include the BBC websites, www.britannica.com, the Chemicool Periodic Table at www.chemicool.com, www.chemguide.co.uk, www.nature.com, and www.newscientist.com.

Callery, Sean, and Miranda Smith. *Periodic Table.* New York: Scholastic, 2017.

Emsley, John. *Nature's Building Blocks, New Edition.* Oxford: Oxford University Press, 2011.

Encyclopædia Britannica, 15th edition. Chicago: Encyclopædia Britannica, Inc., 1977.

Gray, Theodore. *The Elements: A Visual Exploration of Every Known Atom in the Universe.* New York: Black Dog & Leventhal Publishers, 2009.

The Holy Bible. Nashville: Thomas Nelson Publishers, 1976.

Housecroft, Catherine E., and Alan G. Sharpe. *Inorganic Chemistry*, 4th edition. London: Pearson, 2012.

Jackson, Tom. *The Elements Book: A Visual Encyclopedia of the Periodic Table.* London: DK, 2017.

Krebs, Robert E. *The History and Use of Our Earth's Chemical Elements: A Reference Guide*, second edition. Santa Barbara, CA: Greenwood Press, 2006.

Mileham, Rebecca. *The Elements Bible.* Richmond Hill, ON: Firefly Books, 2018.

Padmanabhan, Thanu. *Theoretical Astrophysics: Volume 2, Stars and Stellar Systems.* Cambridge: Cambridge University Press, 2001.

Parsons, Paul, and Gail Dixon. *The Periodic Table: A Visual Guide to the Elements.* London: Quercus, 2014.

Shakespeare, William. *The Complete Works of William Shakespeare.* London: Ramboro Books, 1993.

Weinberg, Steven. *The First Three Minutes: A Modern View of the Origin of the Universe.* New York: Basic Books, 1993.

ABOUT THE AUTHOR

Mary Soon Lee was born and raised in London but has lived in Pittsburgh for over twenty years. She writes both fiction and poetry and has won the Rhysling Award and the Elgin Award. Her work has appeared in a wide range of venues, including *Atlanta Review*, *Cider Press Review*, *American Scholar*, *Science*, *Spillway*, and *Rosebud*, as well as award-winning science fiction and fantasy magazines (*Analog*, *Asimov's*, *Fantasy & Science Fiction*, *Strange Horizons*). Before she took up writing, she had a more analytical background, with degrees in mathematics and computer science from Cambridge University and an MSc in astronautics and space engineering from Cranfield University. She has an antiquated website at www.marysoonlee.com and tweets at @MarySoonLee.

Published in the United States by Ten Speed Press,
an imprint of RandomHouse, a division
of Penguin Random House LLC, New York.
www.tenspeed.com

The haikus were originally published in slightly different form in the
August 4, 2017, issue of *Science.*

Library of Congress Cataloging-in-Publication Data
Names: Lee, Mary Soon, author.
Title: Elemental haiku : poems to honor the periodic table,
three lines at a time / Mary Soon Lee.
Other titles: Poetic periodic table
Description: First edition. | New York : Ten Speed Press, [2019]
Identifiers: LCCN 2019002920 | ISBN 9781984856630 (paperback) |
ISBN 9781984856647 (ebook)
Subjects: LCSH: Periodic table of the elements—Popular works. |
Science—Poetry. | Haiku.
Classification: LCC QD467 .L44 2019 | DDC 546/.8—dc23 LC
record available at https://lccn.loc.gov/2019002920

Trade Paperback ISBN: 978-1-9848-5663-0
eBook ISBN: 978-1-9848-5664-7

Printed in the United States of America

Design by Lisa Schneller Bieser

10 9 8 7 6 5 4 3 2 1

First Edition